Fractions & Decimals
GRADE 4

**Written by
Pam Higdon**

**Illustrated by
Gioia Fiammenghi**

**Cover Illustration
by Susan Cumbow**

FS-12015 Fractions & Decimals Grade 4
All rights reserved—Printed in the U.S.A.

Copyright © 1998 Frank Schaffer Publications, Inc.
23740 Hawthorne Blvd.
Torrance, CA 90505

TABLE OF CONTENTS

Introduction 2
Watermelon Fractions (Parts of a whole) 3
Volcano Fractions (Parts of a set) 4
Peas in a Pod (Comparing fractions) 5
Equal Buckets (Equal fractions) 6
Lemons and Lemonade (Recognizing and writing mixed numbers) 7
A Cookie Recipe (Recognizing and writing fractions and mixed numbers) 8
Snakes End to End (Fractions in measurement) 9
Finish Line (Fractions in measurement) 10
Writing Instruments (Fractional parts) 11
Drawing Stars (Fractional parts) 12
Wrapping Up Fractions (Applications—Fractions) 13
Starfish Fun (Adding fractions with like denominators) 14
What a Mix Up (Adding mixed numbers with like denominators) 15
A Riddle (Subtracting fractions with like denominators) 16
Who Can Subtract? (Subtracting mixed numbers with
    like denominators) 17
Up, Up, and Away (Adding and subtracting fractions and mixed
    numbers with like denominators) 18
Hippity Hoppity (Adding fractions with different denominators) 19
Fraction Fun (Adding fractions with different denominators) 20
Find the Fraction (Subtracting fractions with different denominators) 21
Number Line Fun (Subtracting fractions with different denominators) 22
Let's Practice (Adding and subtracting fractions with
    different denominators) 23
Dots and Bars (Tenths) 24
Where Is the Animal? (Decimals in measurement) 25
Dots and Squares (Hundredths) 26
Climb the Ladders (Adding decimals) 27
The Clock Shop (Adding decimals using money and
    comparing decimals) 28
Paint Shop (Subtracting decimals) 29
Let's Go to Lunch (Subtracting decimals) 30
Answers 31–32

INTRODUCTION

This book is part of a *Math Minders* series. It provides students with opportunities to learn by doing basic skills that will help them understand mathematical properties they will use throughout their lives.

There is no secret to success in math. Children who do well have learned to relate it to their everyday lives, giving math meaning and their study a purpose. As important, however, is the way in which they learn math—one step at a time, with practice. The activities in this series have been created to help students to build upon their understanding of fractions and decimals. Vocabulary for this book is kept at a fourth-grade level to help ensure student success.

A variety of fun formats are included throughout the book to help maintain student interest. Students will be able to progress at their own speed, using the skill learned in one activity to advance to the next step of understanding. The progression is gradual, giving the student a constant feeling of success. The skills and concepts in this book can be taught in the classroom or at home. Some skills covered in this book are fractional parts, adding and subtracting fractions, and adding and subtracting decimals.

Fractions & Decimals

GRADE 4

Name_____

Answer the questions about the watermelons.

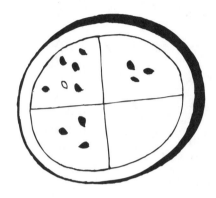

A. What fraction of the watermelon has seeds? _____

B. What fraction of the watermelon does not have seeds? _____

C. What fraction of the seeds are white? _____

D. What fraction of the seeds are dark? _____

E. What fraction of the watermelon has three seeds? _____

F. What fraction of the watermelon has six seeds? _____

G. If Zach, Nicky, Amy, and Stacey each want a piece of watermelon, what fraction of the watermelon can each have? _____

H. What fraction of the watermelons have seeds? _____

I. What fraction of the watermelons do not have seeds? _____

J. What fraction of the seeds are dark? _____

K. What fraction of the seeds are white? _____

L. What fraction of the watermelons have white seeds? _____

M. What fraction of the watermelons have dark seeds? _____

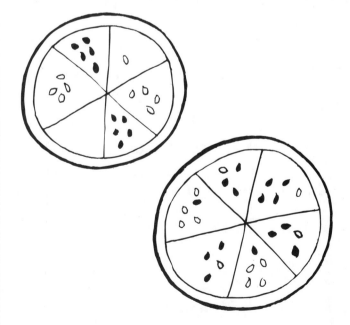

Name_____

Volcano Fractions .... Parts of a set

Use the volcanoes to answer each question.

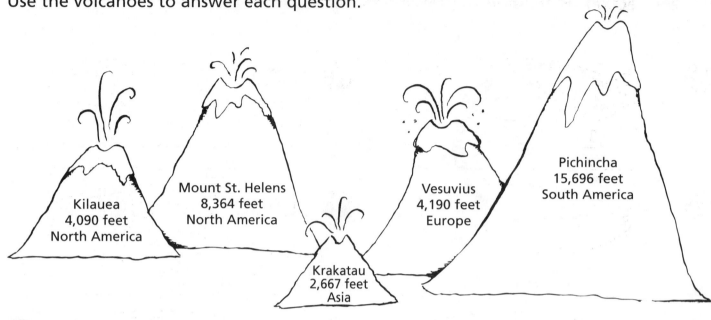

A. The tallest volcano is _____. Write the fraction that tells how often the letter "c" appears in the name. _____

B. Write the fraction that tells how often the letter "i" appears in the name of the tallest volcano. _____

C. Write the fraction that tells how many of the volcanoes are lower than 5,000 feet. _____

D. Write the fraction that tells how many volcanoes are taller than 8,000 feet. _____

E. The shortest volcano is _____. Write the fraction that tells how often the letter "a" appears in the name. _____

F. What fraction of the volcanoes start with the letter "k"? _____

G. What fraction of the volcanoes end with the letter "a"? _____

H. What fraction of the volcanoes are in North America? _____

I. Write the fraction that tells how many of the volcanoes are in Europe. _____

Name_____

Peas in a Pod

Write a fraction for the shaded part of each figure. Then compare the fractions.
Use < or >.

 A.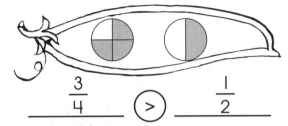

$$\frac{3}{4} \quad > \quad \frac{1}{2}$$

 B.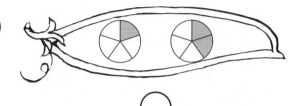

_____ ◯ _____

 C.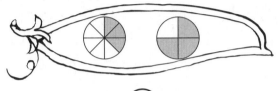

_____ ◯ _____

 D.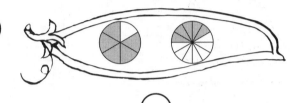

_____ ◯ _____

 E.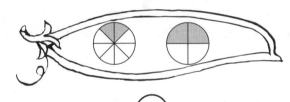

_____ ◯ _____

 F.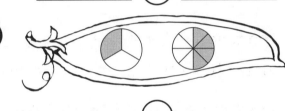

_____ ◯ _____

 G.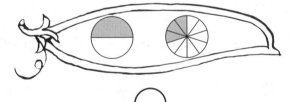

_____ ◯ _____

 H.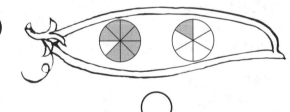

_____ ◯ _____

 I.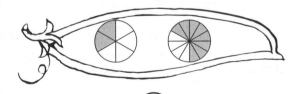

_____ ◯ _____

 J.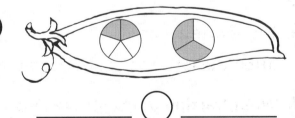

_____ ◯ _____

 K.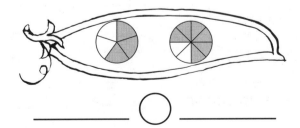

_____ ◯ _____

 L.

_____ ◯ _____

Equal Buckets ....

Use the fraction bars to find the equal fractions.

	$\frac{1}{2}$			$\frac{1}{2}$	
$\frac{1}{3}$		$\frac{1}{3}$		$\frac{1}{3}$	
$\frac{1}{4}$		$\frac{1}{4}$	$\frac{1}{4}$		$\frac{1}{4}$
$\frac{1}{5}$	$\frac{1}{5}$	$\frac{1}{5}$	$\frac{1}{5}$		$\frac{1}{5}$
$\frac{1}{6}$	$\frac{1}{6}$	$\frac{1}{6}$	$\frac{1}{6}$	$\frac{1}{6}$	$\frac{1}{6}$
$\frac{1}{8}$ $\frac{1}{8}$ $\frac{1}{8}$ $\frac{1}{8}$		$\frac{1}{8}$ $\frac{1}{8}$ $\frac{1}{8}$ $\frac{1}{8}$			
$\frac{1}{9}$ $\frac{1}{9}$ $\frac{1}{9}$ $\frac{1}{9}$ $\frac{1}{9}$ $\frac{1}{9}$ $\frac{1}{9}$ $\frac{1}{9}$ $\frac{1}{9}$					
$\frac{1}{10}$ $\frac{1}{10}$ $\frac{1}{10}$ $\frac{1}{10}$ $\frac{1}{10}$ $\frac{1}{10}$ $\frac{1}{10}$ $\frac{1}{10}$ $\frac{1}{10}$ $\frac{1}{10}$					
$\frac{1}{12}$ $\frac{1}{12}$ $\frac{1}{12}$ $\frac{1}{12}$ $\frac{1}{12}$ $\frac{1}{12}$ $\frac{1}{12}$ $\frac{1}{12}$ $\frac{1}{12}$ $\frac{1}{12}$ $\frac{1}{12}$ $\frac{1}{12}$					

A.  $\frac{1}{3}$ =  $\frac{2}{6}$ $\frac{6}{8}$ =  $\frac{}{4}$  $\frac{10}{10}$ = $\frac{}{6}$

B. $\frac{3}{6}$ =  $\frac{}{12}$ 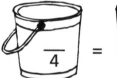 $\frac{}{4}$ =  $\frac{4}{8}$  $\frac{3}{9}$ = $\frac{}{3}$

C.  $\frac{9}{12}$ = $\frac{}{4}$ $\frac{12}{12}$ = $\frac{}{10}$  $\frac{}{12}$ = $\frac{5}{6}$

D.  $\frac{}{8}$ = $\frac{1}{4}$ $\frac{}{5}$ =  $\frac{4}{10}$ $\frac{}{3}$ = $\frac{4}{6}$

E.  $\frac{3}{6}$ = 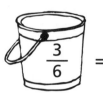 $\frac{}{2}$  $\frac{}{6}$ = $\frac{2}{3}$ $\frac{}{9}$ = $\frac{4}{6}$

Name_____

Lemons and Lemonade

Recognizing and writing mixed numbers

two and one-half glasses

$2\frac{1}{2}$ glasses

one and one-half lemons

$1\frac{1}{2}$ lemons

Write the mixed number.

A. six and seven-eighths $6\frac{7}{8}$

B. two and one-half _____

C. ten and two-thirds _____

D. eight and six-tenths _____

E. one and three-fourths _____

F. nineteen and one-sixth _____

G. three and six-twelfths _____

A mixed number is made up of a whole number and a fraction.

Write a mixed number for each set of pictures.

H. _____

I. _____

J. _____

K. _____

L. _____

FS-12015 Fractions & Decimals Grade 4

Name_____

A Cookie Recipe ......

Use the pictures to write the recipe for chocolate chip cookies.

A. _____ cups flour

B. _____ baking soda

C. _____ teaspoon salt

D. _____ sticks of butter

E. _____ cup white sugar

F. _____ cup brown sugar

G. _____ eggs

H. _____ teaspoons vanilla

I. _____ teaspoon water

J. _____ cups chocolate chips

Name_____

Write the length of each snake.

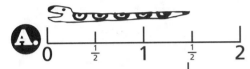

A.
0    $\frac{1}{2}$ 1 $\frac{1}{2}$ 2

This snake is ___$1\frac{1}{2}$___ inches long.

B.
0    $\frac{1}{2}$ 1 $\frac{1}{2}$ 2 $\frac{1}{2}$ 3

This snake is _____ inches long.

C.
0   $\frac{1}{4}$ $\frac{1}{2}$ $\frac{3}{4}$ 1 $\frac{1}{4}$ $\frac{1}{2}$ $\frac{3}{4}$ 2 $\frac{1}{4}$ $\frac{1}{2}$ $\frac{3}{4}$ 3 $\frac{1}{4}$ $\frac{1}{2}$ $\frac{3}{4}$ 4

This snake is _____ inches long.

D.
0   $\frac{1}{4}$ $\frac{1}{2}$ $\frac{3}{4}$ 1

This snake is _____ inches long.

E.
0   $\frac{1}{4}$ $\frac{1}{2}$ $\frac{3}{4}$ 1 $\frac{1}{4}$ $\frac{1}{2}$ $\frac{3}{4}$ 2 $\frac{1}{4}$ $\frac{1}{2}$ $\frac{3}{4}$ 3 $\frac{1}{4}$ $\frac{1}{2}$ $\frac{3}{4}$ 4 $\frac{1}{4}$ $\frac{1}{2}$ $\frac{3}{4}$ 5 $\frac{1}{4}$ $\frac{1}{2}$ $\frac{3}{4}$ 6

This snake is _____ inches long.

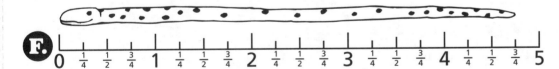

F.
0   $\frac{1}{4}$ $\frac{1}{2}$ $\frac{3}{4}$ 1 $\frac{1}{4}$ $\frac{1}{2}$ $\frac{3}{4}$ 2 $\frac{1}{4}$ $\frac{1}{2}$ $\frac{3}{4}$ 3 $\frac{1}{4}$ $\frac{1}{2}$ $\frac{3}{4}$ 4 $\frac{1}{4}$ $\frac{1}{2}$ $\frac{3}{4}$ 5

This snake is _____ inches long.

G.
0   $\frac{1}{4}$ $\frac{1}{2}$ $\frac{3}{4}$ 1 $\frac{1}{4}$ $\frac{1}{2}$ $\frac{3}{4}$ 2 $\frac{1}{4}$ $\frac{1}{2}$ $\frac{3}{4}$ 3 $\frac{1}{4}$ $\frac{1}{2}$ $\frac{3}{4}$ 4 $\frac{1}{4}$ $\frac{1}{2}$ $\frac{3}{4}$ 5 $\frac{1}{4}$ $\frac{1}{2}$ $\frac{3}{4}$ 6 $\frac{1}{4}$ $\frac{1}{2}$ $\frac{3}{4}$ 7

This snake is _____ inches long.

Name_____

Finish Line ....

See how far each animal traveled in this race. Use your inch ruler to find the length of each animal's path.

 START

 FINISH

A. - - - - - - - - - - - -

The frog jumped _____ inches.

B. - - - - - - - - - - - - -

The penguin wobbled _____ inches.

C. - - - - - - - - - - -

The sloth wandered _____ inches.

D. -

The lizard slithered _____ inches.

E. -

The cheetah bolted _____ inches.

F. - - - - - - - - - - - - - - - - -

The camel lumbered _____ inches.

G. -

The crocodile glided _____ inches.

H. - - - - -

The turtle poked along for _____ inch.

FS-12015 Fractions & Decimals Grade 4

Name_____

Writing Instruments....................

Look at each set of pictures. Circle the fractional part. Then write the fractional part.

A. $\frac{1}{2}$ of 12 = ___6___

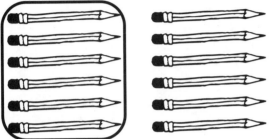

B. $\frac{1}{9}$ of 27 = _____

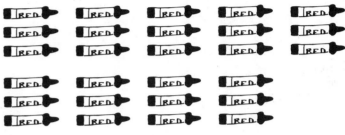

C. $\frac{3}{4}$ of 24 = _____

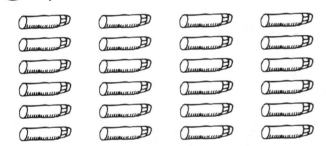

D. $\frac{2}{3}$ of 12 = _____

E. $\frac{3}{10}$ of 20 = _____

F. $\frac{3}{5}$ of 20 = _____

G. $\frac{5}{6}$ of 18 = _____

H. $\frac{4}{7}$ of 28 = _____

   FS-12015 Fractions & Decimals Grade 4

Drawing Stars .... Fractional parts

Find the fractional part. Draw ★'s to help you.

A. $\frac{1}{4}$ of 20 = __5__

★★★★★ ★★★★★
★★★★★ ★★★★★

B. $\frac{1}{3}$ of 18 = _____

C. $\frac{1}{4}$ of 12 = _____

D. $\frac{3}{4}$ of 16 = _____

E. $\frac{1}{5}$ of 15 = _____

F. $\frac{2}{7}$ of 21 = _____

G. $\frac{3}{8}$ of 16 = _____

H. $\frac{2}{6}$ of 18 = _____

I. $\frac{2}{4}$ of 20 = _____

J. $\frac{1}{9}$ of 18 = _____

FS-12015 Fractions & Decimals Grade 4

Name_____

Write a fraction to show how much of each gift is shaded.

A.
_____

B.
_____

C.
_____

D.
_____

E. What fraction of the gifts are ?
_____

F. What fraction of the gifts are ?
_____

Compare the fractions. Use < or >.

G. $\dfrac{2}{3}$ ◯ $\dfrac{2}{5}$

H. $\dfrac{5}{5}$ ◯ $\dfrac{2}{5}$

I. $\dfrac{6}{10}$ ◯ $\dfrac{2}{5}$

J. $\dfrac{4}{16}$ ◯ $\dfrac{7}{16}$

Use the fraction strips to find equal fractions.

K. $\dfrac{3}{4} = \dfrac{\text{☐}}{8}$

L. $\dfrac{\text{☐}}{6} = \dfrac{1}{2}$

M. $\dfrac{1}{2} = \dfrac{\text{☐}}{4}$

$\frac{1}{2}$				$\frac{1}{2}$			
$\frac{1}{4}$		$\frac{1}{4}$		$\frac{1}{4}$		$\frac{1}{4}$	
$\frac{1}{6}$	$\frac{1}{6}$	$\frac{1}{6}$		$\frac{1}{6}$	$\frac{1}{6}$	$\frac{1}{6}$	
$\frac{1}{8}$	$\frac{1}{8}$	$\frac{1}{8}$	$\frac{1}{8}$	$\frac{1}{8}$	$\frac{1}{8}$	$\frac{1}{8}$	$\frac{1}{8}$

Write a mixed number.

N. _____

O. _____

13

Starfish Fun .... Adding fractions with like denominators

Add. Remember that you add the numerators when the denominators are the same.

A. $\frac{2}{6} + \frac{3}{6} = \frac{5}{6}$

B. $\frac{5}{8} + \frac{2}{8} =$

C. $\frac{7}{16} + \frac{5}{16} =$

D. $\frac{9}{20} + \frac{5}{20} =$

E. $\frac{6}{18} + \frac{5}{18} =$

F. $\frac{4}{11} + \frac{6}{11} =$

G. $\frac{2}{7} + \frac{1}{7} =$

H. $\frac{2}{15} + \frac{6}{15} =$

I. $\frac{3}{9} + \frac{2}{9} =$

J. $\frac{4}{12} + \frac{3}{12} =$

K. $\frac{10}{14} + \frac{2}{14} =$

L. $\frac{5}{16} + \frac{8}{16} =$

M. $\frac{3}{10} + \frac{5}{10} =$

N. $\frac{3}{25} + \frac{4}{25} =$

O. $\frac{9}{28} + \frac{4}{28} =$

P. $\frac{1}{5} + \frac{1}{5} =$

What a Mix Up........................
Adding mixed numbers with like denominators

> **Steps to add mixed numbers:**
> 1. Add the fractions.
> 2. Add the whole numbers.

Add.

A. $1\frac{5}{8} + 1\frac{1}{8} =$ _____

B. $8\frac{1}{10} + 2\frac{5}{10} =$ _____

C. $4\frac{7}{16} + 1\frac{6}{16} =$ _____

D. $2\frac{1}{3} + 3\frac{1}{3} =$ _____

E. $7\frac{7}{25} + 8\frac{9}{25} =$ _____

F. $1\frac{1}{5} + 2\frac{2}{5} =$ _____

G. $10\frac{1}{9} + 4\frac{6}{9} =$ _____

H. $3\frac{3}{15} + 6\frac{4}{15} =$ _____

I. $2\frac{1}{6} + 5\frac{2}{6} =$ _____

J. $\begin{array}{r} 6\frac{12}{20} \\ + 6\frac{2}{20} \\ \hline \end{array}$

K. $\begin{array}{r} 4\frac{1}{7} \\ + 5\frac{4}{7} \\ \hline \end{array}$

L. $\begin{array}{r} 3\frac{1}{4} \\ + 8\frac{2}{4} \\ \hline \end{array}$

M. $\begin{array}{r} 4\frac{2}{9} \\ + 1\frac{5}{9} \\ \hline \end{array}$

N. $\begin{array}{r} 9\frac{5}{12} \\ + 1\frac{2}{12} \\ \hline \end{array}$

O. $\begin{array}{r} 6\frac{3}{18} \\ + 2\frac{6}{18} \\ \hline \end{array}$

P. $\begin{array}{r} 7\frac{4}{14} \\ + 2\frac{1}{14} \\ \hline \end{array}$

Q. $\begin{array}{r} 12\frac{6}{18} \\ + 3\frac{9}{18} \\ \hline \end{array}$

R. $\begin{array}{r} 3\frac{7}{16} \\ + 4\frac{5}{16} \\ \hline \end{array}$

S. $\begin{array}{r} 3\frac{5}{8} \\ + 6\frac{2}{8} \\ \hline \end{array}$

T. $\begin{array}{r} 2\frac{2}{11} \\ + 3\frac{6}{11} \\ \hline \end{array}$

U. $\begin{array}{r} 3\frac{2}{14} \\ + 5\frac{3}{14} \\ \hline \end{array}$

FS-12015 Fractions & Decimals Grade 4

A Riddle...................

Subtract. Remember to subtract the numerators when the denominators are the same.

A. $\dfrac{7}{8} - \dfrac{5}{8} = $ _____

B. $\dfrac{9}{10} - \dfrac{3}{10} = $ _____

C. $\dfrac{14}{15} - \dfrac{5}{15} = $ _____

D. $\dfrac{12}{16} - \dfrac{9}{16} = $ _____

E. $\dfrac{8}{9} - \dfrac{4}{9} = $ _____

F. $\dfrac{2}{3} - \dfrac{1}{3} = $ _____

G. $\dfrac{9}{11} - \dfrac{5}{11} = $ _____

H. $\dfrac{3}{4} - \dfrac{1}{4} = $ _____

I. $\dfrac{9}{13} - \dfrac{7}{13} = $ _____

J. $\dfrac{3}{5} - \dfrac{2}{5} = $ _____

K. $\dfrac{8}{12} - \dfrac{5}{12} = $ _____

L. $\dfrac{5}{6} - \dfrac{3}{6} = $ _____

M. $\dfrac{4}{7} - \dfrac{3}{7} = $ _____

N. $\dfrac{10}{14} - \dfrac{6}{14} = $ _____

O. $\dfrac{9}{10} - \dfrac{7}{10} = $ _____

P. $\dfrac{12}{15} - \dfrac{6}{15} = $ _____

Q. $\dfrac{14}{18} - \dfrac{7}{18} = $ _____

R. $\dfrac{15}{20} - \dfrac{8}{20} = $ _____

S. $\dfrac{10}{16} - \dfrac{4}{16} = $ _____

T. $\dfrac{6}{12} - \dfrac{1}{12} = $ _____

U. $\dfrac{6}{7} - \dfrac{2}{7} = $ _____

V. $\dfrac{3}{5} - \dfrac{1}{5} = $ _____

W. $\dfrac{9}{9} - \dfrac{2}{9} = $ _____

X. $\dfrac{8}{11} - \dfrac{3}{11} = $ _____

Y. $\dfrac{9}{13} - \dfrac{4}{13} = $ _____

Z. $\dfrac{7}{8} - \dfrac{2}{8} = $ _____

To find the answer to the riddle, use the answers above to find the letters.

Why did the math student bring a ruler to bed?

$\dfrac{2}{4}$ $\dfrac{4}{9}$ $\dfrac{7}{9}$ $\dfrac{2}{8}$ $\dfrac{4}{14}$ $\dfrac{5}{12}$ $\dfrac{4}{9}$ $\dfrac{3}{16}$ $\dfrac{5}{12}$ $\dfrac{2}{10}$ $\dfrac{6}{16}$ $\dfrac{4}{9}$ $\dfrac{4}{9}$ $\dfrac{2}{4}$ $\dfrac{2}{10}$ $\dfrac{7}{9}$

$\dfrac{2}{6}$ $\dfrac{2}{10}$ $\dfrac{4}{14}$ $\dfrac{4}{11}$ $\dfrac{2}{4}$ $\dfrac{4}{9}$ $\dfrac{6}{16}$ $\dfrac{2}{6}$ $\dfrac{4}{9}$ $\dfrac{6}{15}$ $\dfrac{5}{12}$.

Who Can Subtract? .... Subtracting mixed numbers with like denominators

Steps to subtract mixed numbers:
1. Subtract the fractions.
2. Subtract the whole numbers.

Subtract.

A. $4\frac{2}{5}$
$-1\frac{1}{5}$

B. $6\frac{5}{6}$
$-\frac{4}{6}$

C. $5\frac{3}{4}$
$-2\frac{1}{4}$

D. $9\frac{15}{16}$
$-3\frac{9}{16}$

E. $3\frac{13}{28}$
$-1\frac{9}{28}$

F. $4\frac{15}{33}$
$-2\frac{7}{33}$

G. $2\frac{7}{8}$
$-1\frac{4}{8}$

H. $3\frac{3}{12}$
$-1\frac{1}{12}$

I. $8\frac{4}{5}$
$-7\frac{1}{5}$

J. $3\frac{7}{8}$
$-2\frac{7}{8}$

K. $11\frac{4}{9}$
$-2\frac{2}{9}$

L. $15\frac{8}{9}$
$-12\frac{6}{9}$

Subtract.

M. $5\frac{9}{10} - 1\frac{3}{10} = $ _____

N. $7\frac{8}{10} - 3\frac{2}{10} = $ _____

O. $12\frac{7}{16} - 5\frac{4}{16} = $ _____

P. $9\frac{4}{5} - 2\frac{1}{5} = $ _____

Q. $6\frac{5}{8} - 2\frac{2}{8} = $ _____

R. $10\frac{14}{20} - 6\frac{9}{20} = $ _____

S. $10\frac{8}{11} - 7\frac{5}{11} = $ _____

T. $12\frac{15}{18} - 4\frac{12}{18} = $ _____

Name________________________________

Up, Up, and Away.

Adding and subtracting fractions and mixed numbers with like denominators

Add or subtract.

A. $4\frac{1}{5} + 5\frac{3}{5}$

B. $5\frac{7}{10} - 2\frac{3}{10}$

C. $2\frac{6}{20} + 7\frac{9}{20}$

D. $\frac{4}{5} - \frac{3}{5} =$ ______

E. $4\frac{6}{15} - 1\frac{3}{15}$

F. $7\frac{7}{18} + 5\frac{4}{18}$

G. $5\frac{7}{13} - 2\frac{4}{13}$

H. $\frac{1}{4} + \frac{2}{4} =$ ______

I. $7\frac{5}{8} - 4\frac{3}{8}$

J. $4\frac{2}{9} + 1\frac{5}{9}$

K. $8\frac{1}{10} + 3\frac{3}{10}$

L. $\frac{6}{18} + \frac{7}{18} =$ ______

Name_____

Hippity Hoppity ....

Before adding two fractions, both fractions must have the same denominator. Use the number lines to rewrite the fractions. Then add.

A. $\frac{2}{3} + \frac{1}{6} =$

$\frac{4}{6} + \frac{1}{6} = \frac{5}{6}$

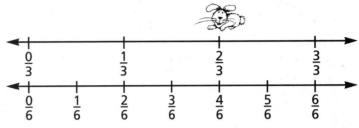

B. $\frac{1}{2} + \frac{1}{6} =$

$\underline{} + \frac{1}{6} = \underline{}$

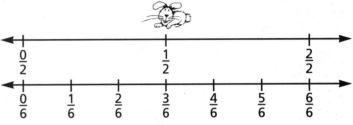

C. $\frac{1}{2} + \frac{1}{4} =$

$\underline{} + \frac{1}{4} = \underline{}$

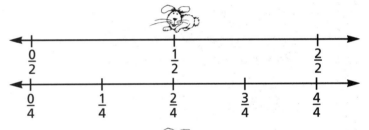

D. $\frac{1}{8} + \frac{1}{2} =$

$\frac{1}{8} + \underline{} = \underline{}$

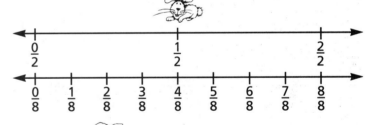

E. $\frac{1}{4} + \frac{1}{8} =$

$\underline{} + \frac{1}{8} = \underline{}$

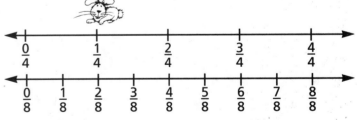

F. $\frac{1}{10} + \frac{1}{5} =$

$\frac{1}{10} + \underline{} = \underline{}$

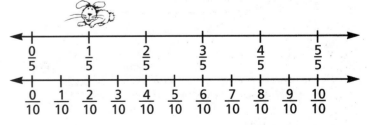

FS-12015 Fractions & Decimals Grade 4

Name_____

Before adding two fractions, both fractions must have the same denominator. Use the number lines to rewrite the fractions. Then add.

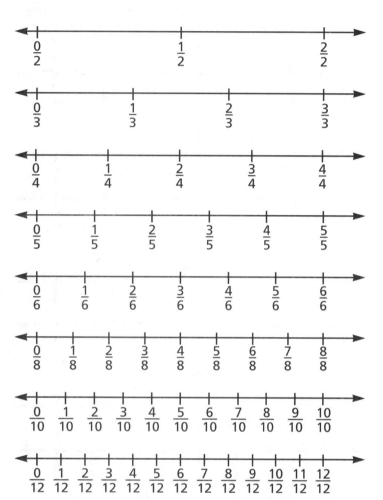

A. $\frac{1}{4} + \frac{5}{8} =$ _____

B. $\frac{1}{2} + \frac{3}{10} =$ _____

C. $\frac{1}{3} + \frac{1}{6} =$ _____

D. $\frac{1}{2} + \frac{3}{12} =$ _____

E. $\frac{1}{2} + \frac{5}{6} =$ _____

F. $\frac{1}{2} + \frac{3}{4} =$ _____

G. $\frac{2}{5} + \frac{3}{10} =$ _____

H. $\frac{9}{10} + \frac{3}{5} =$ _____

I. $\frac{3}{4} + \frac{7}{8} =$ _____

J. $\frac{7}{10} + \frac{3}{5} =$ _____

K. $\frac{2}{3} + \frac{1}{6} =$ _____

L. $\frac{1}{10} + \frac{1}{5} =$ _____

M. $\frac{3}{12} + \frac{2}{3} =$ _____

N. $\frac{2}{4} + \frac{7}{8} =$ _____

O. $\frac{1}{4} + \frac{8}{12} =$ _____

P. $\frac{1}{8} + \frac{3}{4} =$ _____

Q. $\frac{7}{12} + \frac{2}{6} =$ _____

Name_____

Find the Fraction ....

Before subtracting two fractions, both fractions must have the same denominator. Use the number lines to rewrite the fractions. Then subtract.

A. $\dfrac{1}{2} - \dfrac{3}{8} =$

$\dfrac{4}{8} - \dfrac{3}{8} = \dfrac{1}{8}$

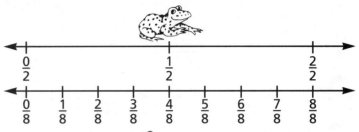

B. $\dfrac{2}{5} - \dfrac{1}{10} =$

$\dfrac{}{} - \dfrac{1}{10} = \dfrac{}{}$

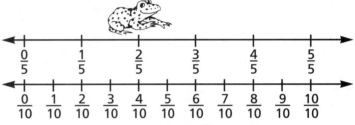

C. $\dfrac{7}{10} - \dfrac{3}{5} =$

$\dfrac{7}{10} - \dfrac{}{} = \dfrac{}{}$

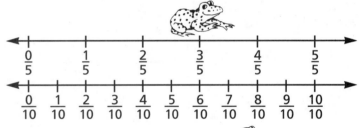

D. $\dfrac{3}{4} - \dfrac{5}{8} =$

$\dfrac{}{} - \dfrac{5}{8} = \dfrac{}{}$

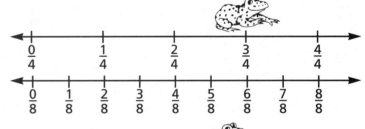

E. $\dfrac{5}{6} - \dfrac{2}{3} =$

$\dfrac{5}{6} - \dfrac{}{} = \dfrac{}{}$

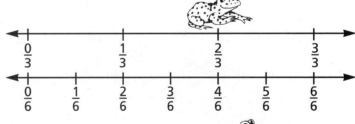

F. $\dfrac{3}{4} - \dfrac{5}{12} =$

$\dfrac{}{} - \dfrac{5}{12} = \dfrac{}{}$

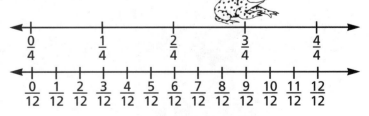

FS-12015 Fractions & Decimals Grade 4

Name_____

Number Line Fun .... Subtracting fractions with different denominators

Use the number lines to rewrite the fractions. Then subtract.

A. $\dfrac{8}{10} - \dfrac{1}{2} = $ $\quad \dfrac{8}{10} - \dfrac{5}{10} = \dfrac{3}{10}$

B. $\dfrac{5}{8} - \dfrac{1}{4} = $ _____

C. $\dfrac{11}{12} - \dfrac{1}{3} = $ _____

D. $\dfrac{7}{10} - \dfrac{3}{5} = $ _____

E. $\dfrac{3}{8} - \dfrac{1}{4} = $ _____

F. $\dfrac{1}{2} - \dfrac{3}{8} = $ _____

G. $\dfrac{1}{2} - \dfrac{1}{12} = $ _____

H. $\dfrac{3}{4} - \dfrac{5}{12} = $ _____

I. $\dfrac{1}{3} - \dfrac{1}{12} = $ _____

J. $\dfrac{3}{5} - \dfrac{3}{10} = $ _____

K. $\dfrac{4}{5} - \dfrac{5}{10} = $ _____

L. $\dfrac{3}{4} - \dfrac{1}{8} = $ _____

M. $\dfrac{1}{3} - \dfrac{1}{6} = $ _____

N. $\dfrac{5}{8} - \dfrac{2}{4} = $ _____

O. $\dfrac{7}{12} - \dfrac{1}{4} = $ _____

P. $\dfrac{9}{10} - \dfrac{2}{5} = $ _____

Q. $\dfrac{2}{3} - \dfrac{1}{6} = $ _____

R. $\dfrac{9}{10} - \dfrac{2}{5} = $ _____

Before subtracting two fractions, both fractions must have the same denominator.

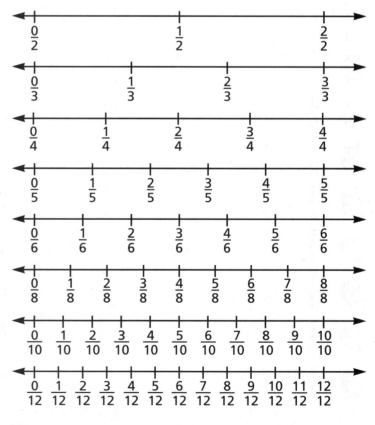

 FS-12015 Fractions & Decimals Grade 4

Name_____

Before adding or subtracting two fractions, both fractions must have the same denominator.

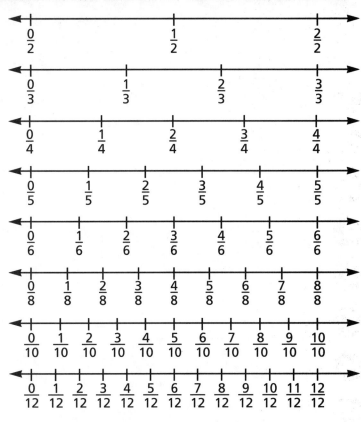

Use the number lines to rewrite the fractions. Then add or subtract.

A. $\frac{1}{4} + \frac{3}{8} =$ $\frac{2}{8} + \frac{3}{8} = \frac{5}{8}$

B. $\frac{1}{2} + \frac{2}{6} =$ _____

C. $\frac{3}{5} - \frac{3}{10} =$ _____

D. $\frac{7}{10} - \frac{3}{5} =$ _____

E. $\frac{1}{2} + \frac{4}{10} =$ _____

F. $\frac{1}{6} + \frac{2}{3} =$ _____

G. $\frac{1}{2} + \frac{3}{8} =$ _____

H. $\frac{5}{8} - \frac{2}{4} =$ _____

I. $\frac{3}{4} - \frac{5}{12} =$ _____

J. $\frac{5}{6} - \frac{1}{12} =$ _____

K. $\frac{3}{4} - \frac{1}{8} =$ _____

L. $\frac{1}{2} + \frac{2}{10} =$ _____

M. $\frac{1}{3} + \frac{1}{6} =$ _____

N. $\frac{1}{4} + \frac{5}{8} =$ _____

O. $\frac{1}{3} - \frac{1}{12} =$ _____

P. $\frac{7}{12} - \frac{1}{4} =$ _____

Name_____

Dots and Bars..........................

$\frac{1}{10}$ or 0.1 is shaded.

When a whole is divided into 10 equal parts, each part is ¹/₁₀ or 0.1.

Write a mixed number and a decimal to show how much is shaded.

A. $1\frac{3}{10}$

1.3

B. _____

_____

C. _____

_____

D. _____

_____

 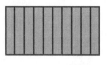 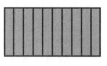 

E. _____

_____

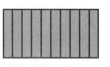  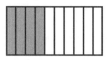 

F. _____

_____

 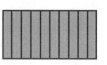  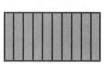 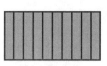 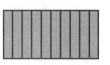 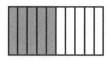 

FS-12015 Fractions & Decimals Grade 4

Name_____

Where Is the Animal? . . . .

Look at the number lines. Write the decimal that tells the position of each animal.

A.

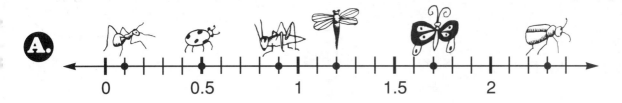

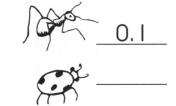   0.1

   _____

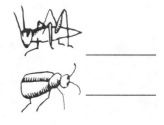   _____

_____   _____

B.

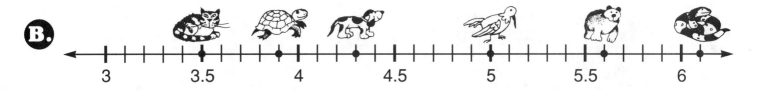

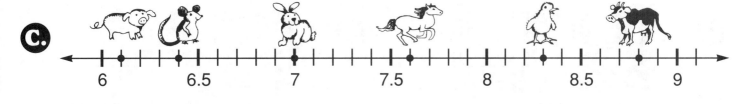

_____   _____   _____

_____   _____   _____

C.

   _____

_____

   _____

_____

   _____

_____

FS-12015 Fractions & Decimals Grade 4

Dots and Squares

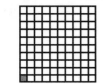

$\frac{1}{100}$ or 0.01 is shaded.

When a whole is divided into 100 equal parts, each part is ¹⁄₁₀₀ or 0.01.

Circle the decimal that tells how much is shaded.

A.

    1.20   (1.60)   1.06

B.

    1.43   1.40   1.45

C.

    1.08   1.80   1.88

D.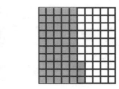

    1.63   1.35   1.53

E.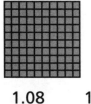

    1.08   1.80   1.88

F.

    1.03   1.30   1.33

G.

    1.01   1.11   1.12

H.

    1.15   1.50   1.51

I.

    1.12   1.20   1.22

J.

    1.04   1.40   1.44

FS-12015 Fractions & Decimals Grade 4

Name_____

Start at the bottom of each ladder. Add until you get to the top.

F. 1.96 + 1.05	**L.** 2.13 + 1.25	**R.** 2.41 + 1.62
E. 0.8 + 1.4	**K.** 1.15 + 4.07	**Q.** 0.81 + 0.05
D. 0.83 + 0.17	**J.** 0.17 + 1.17	**P.** 0.54 + 0.72
C. 0.25 + 0.36	**I.** 1.05 + 0.64	**O.** 1.02 + 0.36
B. 1.2 + 0.9	**H.** 0.9 + 0.4	**N.** 0.4 + 0.5
A. 0.7 + 0.9 = 1.6	**G.** 0.6 + 0.4	**M.** 0.5 + 0.8

FS-12015 Fractions & Decimals Grade 4

Name_____

The Clock Shop ·····················

Find the total price of each pair of clocks. Then compare the total prices. Use < or >.

A.

$56.27
+ $22.36
$71.48

$19.65
+ $45.91
$65.58

(>)

B.

$13.67
+ $12.92

$13.99
+ $12.68

( )

C.

$27.92
+ $16.86

$18.55
+ $25.99

( )

D.

$56.98
+ $25.89

$23.29
+ $58.62

( )

E.

$12.98
+ $15.27

$18.75
+ $13.29

( )

F.

$30.54
+ $12.78

$26.15
+ $21.07

( )

G.

$20.58
+ $13.85

$36.29
+ $10.46

( )

H.

$15.20
+ $22.81

$19.57
+ $18.40

( )

FS-12015 Fractions & Decimals Grade 4

Paint Shop .... Subtracting decimals

Cameron needs to paint the cars. Subtract. Use the chart to color the cars.

0.1 – 2.0	2.1 – 3.0	3.1 – 4.0	4.1 – 5.0	5.1 – 6.0
blue	yellow	orange	green	red

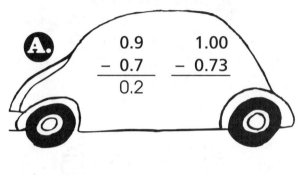

A.

```
  0.9        1.00
- 0.7      - 0.73
-----
  0.2
```

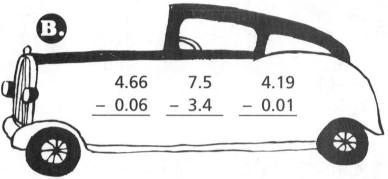

B.

```
  4.66       7.5        4.19
- 0.06     - 3.4      - 0.01
```

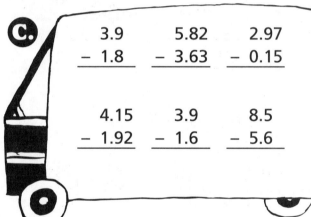

C.

```
  3.9        5.82       2.97
- 1.8      - 3.63     - 0.15


  4.15       3.9        8.5
- 1.92     - 1.6      - 5.6
```

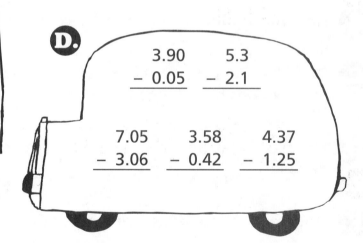

D.

```
  3.90       5.3
- 0.05     - 2.1


  7.05       3.58       4.37
- 3.06     - 0.42     - 1.25
```

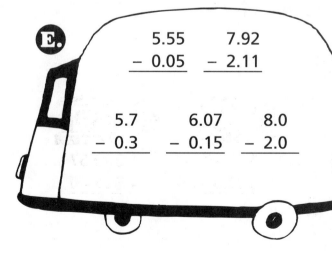

E.

```
  5.55       7.92
- 0.05     - 2.11


  5.7        6.07       8.0
- 0.3      - 0.15     - 2.0
```

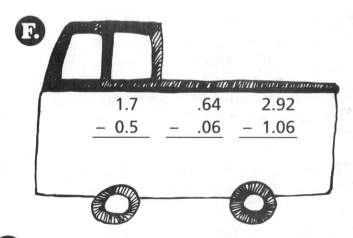

F.

```
  1.7        .64        2.92
- 0.5      - .06      - 1.06
```

FS-12015 Fractions & Decimals Grade 4

Name_____

Let's Go to Lunch ....

Use the prices to solve each problem.

| ham sandwich $1.56 | orange $0.56 | apple $0.92 | cookie $2.43 | hamburger $2.98 | french fries $1.66 |

A. Carissa has $5.00. She buys a hamburger. How much money does Carissa have left?

$$\begin{array}{r} \$\,5.00 \\ -\ \$\,2.98 \\ \hline \$\,2.02 \end{array}$$

B. How much more does an apple cost than an orange? _____

C. Daryl has 57¢. He wants to buy a ham sandwich. How much money does he still need? _____

D. Elsa has $3.00. She buys a cookie. How much money does she have left? _____

E. How much more does a hamburger cost than french fries? _____

F. Shelia has 85¢. She wants to buy some french fries. How much money does Shelia still need? _____

G. Frederick has $1.00. He buys an orange. How much money does he have left? _____

H. Avie has $2.75. He buys a ham sandwich. How much money does he have left? _____

FS-12015 Fractions & Decimals Grade 4

ANSWERS

Page 3
A. $\frac{3}{4}$
B. $\frac{1}{4}$
C. $\frac{1}{12}$
D. $\frac{11}{12}$
E. $\frac{2}{4}$
F. $\frac{1}{4}$
G. $\frac{1}{4}$
H. $1\frac{5}{6}$
I. $\frac{1}{6}$
J. $\frac{25}{46}$
K. $\frac{21}{46}$
L. $1\frac{3}{6}$
M. $1\frac{2}{6}$

Page 4
A. Pichincha; $\frac{2}{9}$
B. $\frac{2}{9}$
C. $\frac{3}{5}$
D. $\frac{2}{5}$
E. Krakatau; $\frac{3}{8}$
F. $\frac{2}{5}$
G. $\frac{2}{5}$
H. $\frac{2}{5}$
I. $\frac{1}{5}$

Page 5
A. $\frac{3}{4} > \frac{1}{2}$
B. $\frac{1}{5} < \frac{2}{5}$
C. $\frac{3}{8} < \frac{3}{4}$
D. $\frac{5}{6} > \frac{5}{12}$
E. $\frac{2}{8} < \frac{2}{4}$
F. $\frac{1}{3} < \frac{4}{8}$
G. $\frac{1}{2} > \frac{3}{10}$
H. $\frac{7}{8} > \frac{1}{6}$
I. $\frac{2}{6} < \frac{8}{12}$
J. $\frac{2}{5} < \frac{2}{3}$
K. $\frac{3}{5} < \frac{6}{8}$
L. $\frac{8}{10} > \frac{5}{10}$

Page 6
A. $\frac{1}{3} = \frac{2}{6}$; $\frac{6}{8} = \frac{3}{4}$; $\frac{10}{10} = \frac{6}{6}$
B. $\frac{3}{6} = \frac{6}{12}$; $\frac{2}{4} = \frac{4}{8}$; $\frac{3}{9} = \frac{1}{3}$
C. $\frac{9}{12} = \frac{3}{4}$; $\frac{12}{12} = \frac{10}{10}$; $\frac{10}{12} = \frac{5}{6}$
D. $\frac{2}{8} = \frac{1}{4}$; $\frac{2}{5} = \frac{4}{10}$; $\frac{2}{3} = \frac{4}{6}$
E. $\frac{3}{6} = \frac{1}{2}$; $\frac{4}{6} = \frac{2}{3}$; $\frac{6}{9} = \frac{4}{6}$

Page 7
A. $6\frac{7}{8}$
B. $2\frac{1}{2}$
C. $10\frac{2}{3}$
D. $8\frac{6}{10}$
E. $1\frac{3}{4}$
F. $19\frac{1}{6}$
G. $3\frac{6}{12}$
H. $5\frac{1}{4}$
I. $2\frac{2}{3}$
J. $3\frac{1}{2}$
K. $3\frac{3}{8}$
L. $4\frac{1}{6}$

Page 8
A. $3\frac{2}{4}$
B. $\frac{2}{4}$
C. $\frac{2}{8}$
D. $1\frac{1}{6}$
E. $\frac{3}{4}$
F. $\frac{2}{4}$
G. $2\frac{1}{2}$
H. $2\frac{1}{4}$
I. $\frac{1}{4}$
J. $1\frac{1}{4}$

Page 9
A. $1\frac{1}{2}$
B. $2\frac{1}{2}$
C. $3\frac{1}{4}$
D. $\frac{3}{4}$
E. $5\frac{1}{4}$
F. $4\frac{3}{4}$
G. $6\frac{1}{2}$

Page 10
A. $1\frac{1}{2}$
B. $2\frac{3}{4}$
C. $1\frac{3}{4}$
D. $4\frac{1}{4}$
E. $5\frac{3}{4}$
F. $3\frac{1}{2}$
G. $4\frac{3}{4}$
H. $\frac{1}{4}$

Page 11
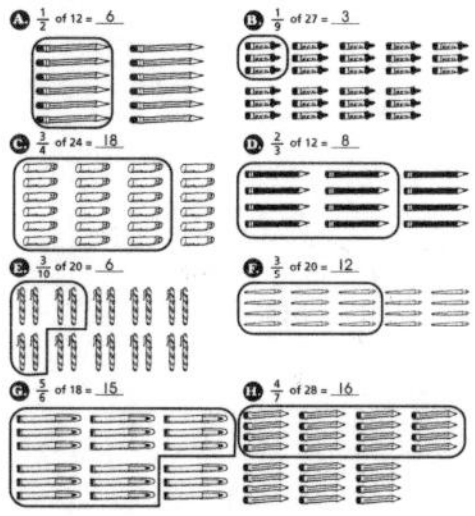
A. 1/2 of 12 = 6
B. 1/9 of 27 = 3
C. 3/4 of 24 = 18
D. 2/3 of 12 = 8
E. 3/10 of 20 = 6
F. 3/5 of 20 = 12
G. 5/6 of 18 = 15
H. 4/7 of 28 = 16

Page 12
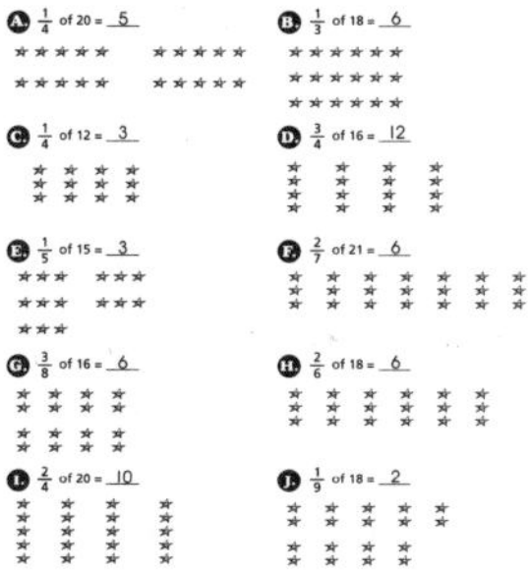
A. 1/4 of 20 = 5
B. 1/3 of 18 = 6
C. 1/4 of 12 = 3
D. 3/4 of 16 = 12
E. 1/5 of 15 = 3
F. 2/7 of 21 = 6
G. 3/8 of 16 = 6
H. 2/6 of 18 = 6
I. 2/4 of 20 = 10
J. 1/9 of 18 = 2

Page 13
A. $\frac{3}{4}$
B. $\frac{5}{8}$
C. $\frac{3}{12}$
D. $\frac{7}{16}$
E. $\frac{2}{5}$
F. $\frac{3}{8}$
G. $>$
H. $>$
I. $>$
J. $<$
K. $\frac{3}{4} = \frac{6}{8}$
L. $\frac{3}{6} = \frac{1}{2}$
M. $\frac{1}{2} = \frac{2}{4}$
N. $2\frac{6}{8}$
O. $4\frac{1}{4}$

Page 14
A. $\frac{5}{6}$
B. $\frac{7}{8}$
C. $\frac{12}{16}$
D. $\frac{14}{20}$
E. $\frac{11}{18}$
F. $\frac{10}{11}$
G. $\frac{3}{7}$
H. $\frac{8}{15}$
I. $\frac{5}{9}$
J. $\frac{7}{12}$
K. $\frac{12}{14}$
L. $\frac{13}{16}$
M. $\frac{8}{10}$
N. $\frac{7}{25}$
O. $\frac{13}{28}$
P. $\frac{2}{5}$

Page 15
A. $2\frac{6}{8}$
B. $10\frac{6}{10}$
C. $5\frac{13}{16}$
D. $5\frac{2}{3}$
E. $15\frac{16}{25}$
F. $3\frac{3}{5}$
G. $14\frac{7}{9}$
H. $9\frac{7}{15}$
I. $7\frac{3}{6}$
J. $12\frac{14}{20}$
K. $9\frac{5}{7}$
L. $11\frac{3}{4}$
M. $5\frac{7}{9}$
N. $10\frac{7}{12}$
O. $8\frac{9}{18}$
P. $9\frac{5}{14}$
Q. $15\frac{15}{18}$
R. $7\frac{12}{16}$
S. $9\frac{7}{8}$
T. $5\frac{8}{11}$
U. $8\frac{5}{14}$

Page 16
A. $\frac{2}{8}$
B. $\frac{6}{10}$
C. $\frac{9}{15}$
D. $\frac{3}{16}$
E. $\frac{4}{9}$
F. $\frac{1}{3}$
G. $\frac{4}{11}$
H. $\frac{2}{4}$
I. $\frac{2}{13}$
J. $\frac{1}{5}$
K. $\frac{3}{12}$
L. $\frac{2}{6}$
M. $\frac{1}{7}$
N. $\frac{4}{14}$
O. $\frac{2}{10}$
P. $\frac{6}{15}$
Q. $\frac{7}{18}$
R. $\frac{7}{20}$
S. $\frac{6}{16}$
T. $\frac{5}{12}$
U. $\frac{4}{7}$
V. $\frac{2}{5}$
W. $\frac{7}{9}$
X. $\frac{5}{11}$
Y. $\frac{5}{13}$
Z. $\frac{5}{8}$

He wanted to see how long he slept.

Page 17
A. $3\frac{1}{5}$
B. $6\frac{1}{6}$
C. $3\frac{2}{4}$
D. $6\frac{6}{16}$
E. $2\frac{4}{28}$
F. $2\frac{8}{33}$
G. $1\frac{3}{8}$
H. $2\frac{2}{12}$
I. $1\frac{3}{5}$
J. $1\frac{6}{8}$
K. $9\frac{2}{9}$
L. $3\frac{2}{9}$
M. $4\frac{6}{10}$
N. $4\frac{6}{10}$
O. $7\frac{3}{16}$
P. $7\frac{3}{5}$
Q. $4\frac{3}{8}$
R. $4\frac{5}{20}$
S. $3\frac{3}{11}$
T. $8\frac{3}{18}$

Page 18
A. $9\frac{4}{5}$
B. $3\frac{4}{10}$
C. $9\frac{15}{20}$
D. $\frac{1}{5}$
E. $3\frac{3}{15}$
F. $12\frac{11}{18}$
G. $3\frac{3}{13}$
H. $\frac{3}{4}$
I. $3\frac{2}{8}$
J. $5\frac{7}{9}$
K. $11\frac{4}{10}$
L. $\frac{13}{18}$

ANSWERS

Page 19
A. $\frac{4}{6} + \frac{1}{6} = \frac{5}{6}$
B. $\frac{3}{6} + \frac{1}{6} = \frac{4}{6}$
C. $\frac{2}{4} + \frac{1}{4} = \frac{3}{4}$
D. $\frac{1}{8} + \frac{4}{8} = \frac{5}{8}$
E. $\frac{2}{8} + \frac{1}{8} = \frac{3}{8}$
F. $\frac{1}{10} + \frac{2}{10} = \frac{3}{10}$

Page 20
A. $\frac{2}{8} + \frac{5}{8} = \frac{7}{8}$
B. $\frac{5}{10} + \frac{3}{10} = \frac{8}{10}$
C. $\frac{2}{6} + \frac{1}{6} = \frac{3}{6}$
D. $\frac{6}{12} + \frac{3}{12} = \frac{9}{12}$
E. $\frac{3}{8} + \frac{5}{8} = \frac{8}{8}$
F. $\frac{2}{4} + \frac{3}{4} = \frac{5}{4}$
G. $\frac{4}{10} + \frac{3}{10} = \frac{7}{10}$
H. $\frac{9}{10} + \frac{6}{10} = \frac{15}{10}$
I. $\frac{6}{8} + \frac{7}{8} = \frac{13}{8}$
J. $\frac{7}{10} + \frac{6}{10} = \frac{13}{10}$
K. $\frac{4}{6} + \frac{1}{6} = \frac{5}{6}$
L. $\frac{1}{10} + \frac{2}{10} = \frac{3}{10}$
M. $\frac{3}{12} + \frac{8}{12} = \frac{11}{12}$
N. $\frac{4}{8} + \frac{7}{8} = \frac{11}{8}$
O. $\frac{3}{12} + \frac{8}{12} = \frac{11}{12}$
P. $\frac{1}{8} + \frac{6}{8} = \frac{7}{8}$
Q. $\frac{7}{12} + \frac{4}{12} = \frac{11}{12}$

Page 21
A. $\frac{4}{8} - \frac{3}{8} = \frac{1}{8}$
B. $\frac{4}{10} - \frac{1}{10} = \frac{3}{10}$
C. $\frac{7}{10} - \frac{6}{10} = \frac{1}{10}$
D. $\frac{6}{8} - \frac{5}{8} = \frac{1}{8}$
E. $\frac{5}{6} - \frac{4}{6} = \frac{1}{6}$
F. $\frac{9}{12} - \frac{5}{12} = \frac{4}{12}$

Page 22
A. $\frac{8}{10} - \frac{5}{10} = \frac{3}{10}$
B. $\frac{5}{8} - \frac{2}{8} = \frac{3}{8}$
C. $\frac{11}{12} - \frac{4}{12} = \frac{7}{12}$
D. $\frac{7}{10} - \frac{6}{10} = \frac{1}{10}$
E. $\frac{3}{8} - \frac{2}{8} = \frac{1}{8}$
F. $\frac{4}{8} - \frac{3}{8} = \frac{1}{8}$
G. $\frac{6}{12} - \frac{1}{12} = \frac{5}{12}$
H. $\frac{9}{12} - \frac{5}{12} = \frac{4}{12}$
I. $\frac{4}{12} - \frac{1}{12} = \frac{3}{12}$
J. $\frac{6}{10} - \frac{3}{10} = \frac{3}{10}$
K. $\frac{8}{10} - \frac{5}{10} = \frac{3}{10}$
L. $\frac{6}{8} - \frac{1}{8} = \frac{5}{8}$
M. $\frac{2}{6} - \frac{1}{6} = \frac{1}{6}$
N. $\frac{5}{8} - \frac{4}{8} = \frac{1}{8}$
O. $\frac{7}{12} - \frac{3}{12} = \frac{4}{12}$
P. $\frac{9}{10} - \frac{4}{10} = \frac{5}{10}$
Q. $\frac{4}{6} - \frac{1}{6} = \frac{3}{6}$
R. $\frac{9}{10} - \frac{4}{10} = \frac{5}{10}$

Page 23
A. $\frac{2}{8} + \frac{3}{8} = \frac{5}{8}$
B. $\frac{3}{6} + \frac{2}{6} = \frac{5}{6}$
C. $\frac{6}{10} - \frac{3}{10} = \frac{3}{10}$
D. $\frac{7}{10} - \frac{6}{10} = \frac{1}{10}$
E. $\frac{5}{10} + \frac{4}{10} = \frac{9}{10}$
F. $\frac{1}{6} + \frac{4}{6} = \frac{5}{6}$
G. $\frac{4}{8} + \frac{3}{8} = \frac{7}{8}$
H. $\frac{5}{8} - \frac{4}{8} = \frac{1}{8}$
I. $\frac{9}{12} - \frac{5}{12} = \frac{4}{12}$
J. $\frac{10}{12} - \frac{1}{12} = \frac{9}{12}$
K. $\frac{6}{8} - \frac{1}{8} = \frac{5}{8}$
L. $\frac{5}{10} + \frac{2}{10} = \frac{7}{10}$
M. $\frac{2}{6} + \frac{1}{6} = \frac{3}{6}$
N. $\frac{2}{8} + \frac{5}{8} = \frac{7}{8}$
O. $\frac{4}{12} - \frac{1}{12} = \frac{3}{12}$
P. $\frac{7}{12} - \frac{3}{12} = \frac{4}{12}$

Page 24
A. $1\frac{3}{10}$; 1.3
B. $2\frac{6}{10}$; 2.6
C. $3\frac{8}{10}$; 3.8
D. $2\frac{4}{10}$; 2.4
E. $4\frac{7}{10}$; 4.7
F. $3\frac{5}{10}$; 3.5

Page 25
A. ant – 0.1
   lady bug – 0.5
   dragonfly – 1.2
   butterfly – 1.7
   grasshopper – 0.9
   beetle – 2.3
B. cat – 3.5
   dog – 4.3
   bird – 5.0
   snake – 6.1
   bear – 5.6
   turtle – 3.9
C. mouse – 6.4
   chick – 8.3
   rabbit – 7.0
   horse – 7.6
   cow – 8.8
   pig – 6.1

Page 26
A. 1.60 B. 1.45
C. 1.08 D. 1.53
E. 1.80 F. 1.30
G. 1.12 H. 1.15
I. 1.22 J. 1.40

Page 27
A. 1.6   G. 1.0   M. 1.3
B. 2.1   H. 1.3   N. 0.9
C. 0.61 I. 1.69 O. 1.38
D. 1.00 J. 1.34 P. 1.26
E. 2.2   K. 5.22 Q. 0.86
F. 3.01 L. 3.38 R. 4.03

Page 28
A. $71.48 > $65.58
B. $26.59 < $26.67
C. $44.78 > $44.54
D. $82.87 > $81.91
E. $28.25 < $32.04
F. $43.32 < $47.22
G. $34.43 < $46.75
H. $38.01 > $37.97

Page 29
A. 0.2; 0.27
   car colored blue
B. 4.60; 4.1; 4.18
   car colored green
C. 2.1; 2.19; 2.82
   2.23; 2.3; 2.9
   car colored yellow
D. 3.85; 3.2
   3.99; 3.16; 3.12
   car colored orange
E. 5.50; 5.81
   5.4; 5.92; 6.0
   car colored red
F. 1.2; .58; 1.86
   car colored blue

Page 30
A. $2.02
B. $0.36
C. $0.99
D. $0.57
E. $1.32
F. $0.81
G. $0.44
H. $1.19

Teachers and Parents!

   Now you can help your children succeed with fractions and decimals. This book will help children with the following skills:

- Comparing fractions
- Adding and subtracting fractions and decimals
- Using decimals in measurement
- And more!

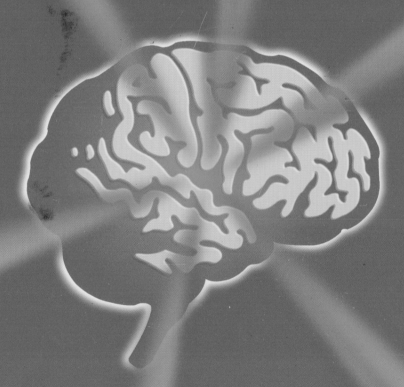

Frank Schaffer Publications, Inc.

CAT. NO. FS-12015
ISBN 0-7682-0031-8
Printed in the U.S.A.